Current Intelligence Bulletin 60

Interim Guidance for Medical Screening and Hazard Surveillance for Workers Potentially Exposed to Engineered Nanoparticles

DEPARTMENT OF HEALTH AND HUMAN SERVICES
Centers for Disease Control and Prevention
National Institute for Occupational Safety and Health

Disclaimer

Mention of any company or product does not constitute endorsement by the National Institute for Occupational Safety and Health (NIOSH). In addition, citations to Web sites external to NIOSH do not constitute NIOSH endorsement of the sponsoring organizations or their programs or products. Furthermore, NIOSH is not responsible for the content of these Web sites. All Web addresses referenced in this document were accessible as of the publication date.

Ordering Information

To receive documents or other information about occupational safety and health topics, contact NIOSH at

Telephone: 1–800–CDC–INFO (1–800–232–4636)
TTY: 1–888–232–6348
E-mail: cdcinfo@cdc.gov

or visit the NIOSH Web site at www.cdc.gov/niosh.

For a monthly update on news at NIOSH, subscribe to *NIOSH eNews* by visiting www.cdc.gov/niosh/eNews.

DHHS (NIOSH) Publication No. 2009–116

February 2009

SAFER • HEALTHIER • PEOPLE™

Foreword

"Do occupational exposures to engineered nanoparticles pose an unintended risk of adverse health effects?" This is not an abstract or theoretical question that practitioners have the luxury of debating for years before it becomes a reality. Nanotechnology is a reality, with potential for great growth in the 21st Century. Workers are already engaged in processes in which they may be exposed to materials that never existed before in nature. We do not fully know how these engineered nanoparticles may enter the body, where they may travel once inside, or what effects they may have on the body's systems. We do not fully know whether or how effects may differ for chemically or structurally different particles at the nanoscale. Diverse stakeholders have agreed that research to address these questions is essential for the responsible development of nanotechnology.

As research progresses to answer those questions, the National Institute for Occupational Safety and Health (NIOSH) has recommended prudent precautionary interim measures for reducing work-related exposures and assessing potential risk. In the hierarchy of prevention, it is important to consider where it may be of value to provide medical screening of workers who may be exposed to a potential health hazard, but who may be asymptomatic—that is, who have no identifiable symptom of an occupational disease. On the frontiers of nanotechnology, where as yet little data exist for assessing risk with confidence, it is difficult to recommend specific screening tests. NIOSH has sought a wide range of opinions on the matter and along with its own review of the scientific literature presents this interim guidance for medical screening and hazard surveillance. The evidence base on the health effects of engineered nanoparticles is rapidly growing and NIOSH will continue to monitor and assess it and will update those recommendations as more definitive information becomes available.

Christine M. Branche, Ph.D.
Acting Director
National Institute for Occupational
    Safety and Health
Centers for Disease Control
    and Prevention

Executive Summary

Concerns have been raised about whether workers exposed to engineered nanoparticles are at increased risk of adverse health effects. The current body of evidence about the possible health risks of occupational exposure to engineered nanoparticles is quite small. While there is increasing evidence to indicate that exposure to some engineered nanoparticles can cause adverse health effects in laboratory animals, no health studies of workers exposed to the few engineered nanoparticles tested in animals have been published. The purpose of this document from the National Institute for Occupational Safety and Health (NIOSH) is to provide interim guidance about whether specific medical screening, including performing medical tests on asymptomatic workers, is appropriate for these workers.

Medical screening is only one part of what should be considered a complete safety and health management program. An ideal safety and health management program follows a hierarchy of controls and involves various occupational health surveillance measures. Since specific medical screening of asymptomatic workers exposed to engineered nanoparticles has not been extensively discussed in the scientific literature, this document makes recommendations based upon what is known until more rigorous research can be performed.

Currently there is **insufficient scientific and medical evidence to recommend the specific medical screening of workers potentially exposed to engineered nanoparticles.** Nonetheless, this lack of evidence does not preclude specific medical screening by employers interested in taking precautions beyond existing industrial hygiene measures. If nanoparticles are composed of a chemical or bulk material for which medical screening recommendations exist, these same screening recommendations would be applicable for workers exposed to engineered nanoparticles as well.

As research into the hazards of engineered nanoparticles continues, vigilant reassessment of available data is critical to determine whether specific medical screening is warranted for workers. **In the interim, the following recommendations are provided for workplaces where workers may be exposed to engineered nanoparticles in the course of their work:**

- **Take prudent measures to control exposures to engineered nanoparticles.**
- **Conduct hazard surveillance as the basis for implementing controls.**
- **Continue use of established medical surveillance approaches.**

NIOSH will continue to collect and evaluate new research findings and update its recommendations about medical screening programs for workers exposed to nanoparticles.

NIOSH will also continue to consider the strengths and weaknesses of establishing exposure registries for workers potentially exposed to engineered nanoparticles for future health surveillance and epidemiological studies.

Contents

Acknowledgements

This Current Intelligence Bulletin (CIB) was developed by the staff of the National Institute for Occupational Safety and Health (NIOSH) who participate in the NIOSH Nanotechnology Research Center (NTRC). Special thanks go to Paul S. Schulte, Director, Education and Information Division, NIOSH and manager of the NTRC, Douglas Trout, and Ralph D. Zumwalde for writing and organizing the report. The NIOSH NTRC also acknowledges the contributions of Vanessa Becks and Gino Fazio for desktop publishing and graphic design, and Douglas Platt for editing the document.

NIOSH greatly appreciates the time and efforts of expert peer reviewers and NTRC staff who provided comments on a draft of this CIB.

Peer Reviewers

Michael Kosnett, MD, MPH
University of Colorado at Denver and Health Sciences Center

Ken Donaldson, BSc, PhD, DSc, CBiol, FRCPath, FFOM
University of Edinburgh Centre for Inflammation Research, ELEGI Colt Laboratory

James Lockey, MD, Professor of Occupational, Environmental and Pulmonary Medicine, University of Cincinnati

Attendees and Participants: Workshop on occupational health surveillance and nanotechnology workers. April 17–18, 2007, Arlington, VA.

Richard Canady FDA	James Collins Dow Chemical	Michael Fischman Intel
Charles Geraci NIOSH/EID	Barbara Gibson 3M	Harold Haase Lockheed Martin
William Halperin UMDNJ	Deanna Harkins U.S. Army CHPPM	John Howard NIOSH/OD
Matt Hull Luna Innovations	Jackie Isaacs NEU, Nano (NSEC)	Amy Jones Lockheed Martin
Steve Joslin Luna Innovations	Anthony Klapper Reed Smith	Michael Kosnett Am Coll Med Tox
Eileen Kuempel NIOSH/EID	Tabitha Maher Altairnano	Robert McCunney MIT

(Continued)

Attendees and Participants: Workshop on occupational health surveillance and nanotechnology workers. April 17–18, 2007, Arlington, VA. (Continued)

James Melius
LIUNA

George Mellendick
Pfizer

Michael Muhm
Boeing

Diane Mundt
Environ International Corp

Kenneth Mundt
Environ International Corp

Vladimir Murashov
NIOSH/OD

Michael Nasterlack
BASF

Minda Nieblas
OSHA

Lyn Penniman
OSHA

John Piacentino
NIOSH/OD

Scott Prothero
EPA

Anita Schill
NIOSH/OD

Mary Schubauer-Berigan
NIOSH/DSHEFS

Paul Schulte
NIOSH/EID

John Sestito
NIOSH/DSHEFS

Clifford Strader
DOE

Pat Sullivan
NIOSH/DRDS

Marie Sweeney
NIOSH/DSHEFS

Douglas Trout
NIOSH/DSHEFS

David Warheit
DuPont

Norbert Will
Clariant

Ralph Zumwalde
NIOSH/EID

Interim Guidance for the Medical Screening and Hazard Surveillance for Workers Potentially Exposed to Engineered Nanoparticles

Introduction

Nanotechnology is a system of innovative methods for controlling and manipulating matter at the near-atomic scale to produce engineered materials, structures, and devices. Engineered nanoparticles are generally considered to include a class or subset of these manufactured materials with at least one dimension of approximately 1 to 100 nanometers (www.nano.gov/html/facts/whatIsNano. html). At these scales, materials often exhibit unique properties beyond those expected at the chemical or bulk level that affect their physical, chemical, and biological behavior. The term "ultrafine" is also frequently used in the literature to describe particles with dimensions less than 100 nanometers that have not been intentionally produced (e.g., manufactured) but are the incidental products of processes involving combustion, welding, or diesel engines. It is currently unclear whether a distinction in particle terminology is justified from a safety and health perspective if the particles have the same physicochemical characteristics.

The potential occupational health risks associated with the manufacture and use of nanomaterials are not yet clearly understood. Many engineered nanomaterials and devices are formed from nanometer-scale particles (i.e., nanoparticles) that are initially produced as aerosols or colloidal suspensions. Exposure to these materials during manufacturing and use may occur through inhalation, dermal contact, or ingestion; however, inhalation exposure is the main route of concern [ASCC 2006]. There is very limited information available about dominant exposure routes, the potential for exposure, and material toxicity.

At this time, society in general and companies in particular are faced with the dilemma of balancing a desire to expand a potentially bountiful technology against the potential hazards that may result. The real risks from the technology are not known, and the perceived risks are undetermined. In this regard, nanotechnology is no different from any other emerging technology. One of the first areas where exposures to engineered nanoparticles will occur is in the workplace. In the face of uncertainty about the hazards of nanoparticles, a corporate or societal response (such as implementing appropriate occupational health surveillance measures) may assure the public that appropriate efforts are being taken to identify and control potential hazards in a timely fashion.

Concerned individuals from government, industry, labor, and academia, together with occupational health professionals and medical personnel, have raised questions about whether workers exposed to engineered nanoparticles should be provided some type of medical surveillance. The purpose of this document is to provide interim guidance concerning specific medical screening for these workers—that is, medical tests for asymptomatic workers—until additional research either supports or negates the need for this type of screening. The type and degree of screening recommended here is in addition to any medical surveillance taking place as part of existing occupational health surveillance efforts.

Background: A Brief Review of the Literature

Effects of Exposure to Ultrafine Particles

Results from epidemiological studies in the general population have shown associations between fine particulate air pollution and increased morbidity and mortality from respiratory and cardiopulmonary disease [Dockery et al. 1993; Ibald-Mulli et al. 2002; Pope et al. 2004]. Other studies have investigated specific markers of effect associated with exposure to the ultrafine particulate fraction of air pollution [Ruckerl et al. 2006]. Studies of workers exposed to ultrafine particles (e.g., diesel exhaust and welding fume) have reported elevated lung cancer risks [Steenland et al. 1998; Garshick et al. 2004; Antonini 2003] while results from some animal studies have shown that many types of poorly soluble ultrafine particles can elicit a greater pulmonary inflammatory response than larger particles of the same composition on a mass for mass basis [Oberdörster et al. 1994; Lison et al 1997; Zhang et al. 2000, 2003; Brown et al. 2001; Höhr et al. 2002; Duffin et al. 2007]. Toxicological studies indicate that the chemical and physical properties that influence the toxicity of ultrafine particles may also be relevant to mechanisms that influence the toxicological response to engineered nanoparticles [Castranova et al. 2000; Aitken et

al. 2004; Donaldson et al. 2005, 2006; Maynard and Kuempel 2005; Oberdörster et al. 2005a, b; Kreyling et al. 2006; Gwinn and Vallyathan 2006; Borm et al. 2006; Helland et al. 2007]. Studies have also shown that physicochemical properties such as surface reactivity, chemical composition, crystal structure, and shape also influence the toxicity of nanoscale particles [Zhang et al. 1998; Dick et al. 2003; Warheit et al. 2007a, b]. Adverse effects reported from exposure to ultrafine particles have raised concerns about workers exposed to engineered nanoparticles [Royal Society and Academy of Engineering 2004; Maynard and Kuempel 2005; IRRST 2006; Nel et al. 2006; Schulte and Salmanca-Buentello 2007; Maynard 2007; Lam et al. 2006; Kuempel et al. 2007; Aitken et al. 2004; ASCC 2006]

Effects of Exposure to Engineered Nanoparticles

Animal studies with some types of engineered nanoparticles have caused adverse lung effects (e.g., pulmonary inflammation and progressive fibrosis) [Lam et al. 2004, 2006; Shvedova et al. 2005; Takagi et al. 2008; Poland et al. 2008] and cardiovascular effects (e.g., inflammation, blood platelet activation, plaque formation, and thrombosis) [Radomski et al. 2005; Donaldson et al. 2006; Li et al. 2007]. Other studies have demonstrated that discrete nanoparticles may enter the bloodstream from the lungs and translocate to other organs [Oberdörster et al. 2002] while other studies have shown that discrete nanoparticles (35-37 nm count median diameter) that deposit in the nasal region may be able to enter the brain by translocation along the olfactory nerve [Oberdörster et al. 2005(b); Elder et al. 2006]. A broader review of the human and animal data can be found in the NIOSH document *Approaches to Safe*

Nanotechnology: An Information Exchange with NIOSH [NIOSH 2006a].

Occupational Health Surveillance

NIOSH has historically recommended implementing occupational health surveillance programs when workers are exposed to potentially hazardous materials. Occupational health surveillance involves the ongoing systematic collection, analysis, and dissemination of exposure and health data on groups of workers for the purpose of preventing illness and injury. This information is frequently used for establishing and evaluating the hierarchy of preventive actions [Halperin 1996]. The general term *occupational health surveillance* includes medical and hazard surveillance. Occupational health surveillance is an essential component of an effective occupational safety and health program [Harber et al. 2003; NIOSH 2006b; Wagner and Fine 2008; Baker and Matte 2005]. While this document supports that concept, *the main focus is whether a typical medical surveillance program that includes additional medical screening is warranted for workers potentially exposed to engineered nanoparticles.*

Medical Surveillance

Medical surveillance targets actual health events or a change in a biologic function of an exposed person or persons. Medical surveillance is a second line of defense behind the implementation of engineering, administrative, and work practice controls including the use of personal protective equipment. NIOSH recommends the medical surveillance of workers when they are exposed to hazardous materials. The elements of a medical surveillance program generally include the following:

1. An initial medical examination and collection of medical and occupational histories.

2. Periodic medical examinations at regularly scheduled intervals, including specific medical screening tests when warranted.

3. More frequent and detailed medical examinations as indicated on the basis of findings from these examinations.

4. Post-incident examinations and medical screening following uncontrolled or nonroutine increases in exposures such as spills.

5. Worker training to recognize symptoms of exposure to a given hazard.

6. A written report of medical findings.

7. Employer actions in response to identification of potential hazards.

When the purpose of a medical surveillance program is to detect early signs of work-related illness and disease, it is considered a type of medical screening, also referred to as medical monitoring and includes medical testing to detect preclinical changes in organ function or changes before a person would normally seek medical care and when intervention is beneficial [Ashford et al. 1990; Baker and Matte 2005; Halperin et al. 1986; Harber et al. 2003; ILO 1998]. The establishment of a medical screening program should follow established criteria [Halperin et al. 1986; Borak et al. 2006; Baker and Matte 2005; Harber 2003] and that specific disease endpoints must be able to be determined by the test selected (see Appendix A).

Frequent Uses for Medical Surveillance

Medical examinations and tests are used in many workplaces to determine whether an employee is able to perform the essential functions of the job, with or without reasonable accommodation, without posing a direct and imminent threat to the safety or health of the worker or others. Workplace medical examinations must be conducted in compliance with the Americans with Disabilities Act of 1990 (ADA) (Public Law No. 101-336). For example, this law prohibits making a job offer contingent upon the applicant's submission to a medical examination. Still, medical examinations and examinations conducted before placing a worker in a given job could potentially provide useful baseline information in a variety of ways. For example, even if there appears to be no reason for immediate concern about exposure to engineered nanoparticles in a particular workplace setting, this type of baseline data may benefit employers and workers alike if questions come up later regarding potential worker health problems associated with the specific engineered nanoparticle.

Medical surveillance of workers is also required by law when there is exposure to a specific workplace hazard. Although OSHA does not have a standard that specifically addresses occupational exposure to engineered nanoparticles, OSHA has a number of standards that require medical surveillance of workers. Workplaces with engineered nanoparticles comprised of chemicals addressed by current OSHA standards (Appendix B) are subject to the requirements of those standards, including the requirements for medical surveillance. In addition, medical surveillance of workers handling engineered nanoparticles may also be triggered when workers are exposed to other hazardous substances (e.g., those listed in Appendix B) present in nanoparticle operations.

In addition to substance-specific standards, OSHA has standards with broader applicability. For example, employers must follow the medical evaluation requirements of OSHA's respiratory protection standard [29 CFR 1910.134] when respirators are necessary to protect worker health. This standard includes elements of medical surveillance. Likewise, the OSHA standard for occupational exposure to hazardous chemicals in laboratories [29 CFR 1910.1450] requires medical consultation following the accidental release of hazardous chemicals.

NIOSH also recommends medical surveillance (including screening) of workers when there is exposure to certain occupational hazards (Appendix C). None of the hazards noted in Appendix C are identified as engineered nanoparticles, but medical surveillance would apply to workers exposed to nanoparticles comprised of chemicals for which NIOSH has a recommendation. The medical surveillance of these workers may provide useful information if questions arise in the future about the health effects of their exposure to nanoparticles.

Hazard Surveillance and Risk Management

Hazard surveillance involves identifying potentially hazardous practices or exposures in the workplace and assessing the extent to which they can be linked to workers, the effectiveness of controls, and the reliability of exposure measures [Sundin and Frazier 1989; Froines et al. 1989]. **Hazard surveillance for engineered nanoparticles is an essential component of any occupational health surveillance effort and is used for defining the elements of the risk management program.** One component of a risk management program involves taking action to minimize exposure to potential hazards. In the case of engineered nanoparticles, even in the absence of adequate health information, an understanding of potential worker exposures can form the basis for ongoing risk management. Other critical elements of a risk management program include recognizing potential exposures and determining appropriate actions to minimize them (e.g., implementing engineering controls, employing good work practices, and using personal protective equipment) [NIOSH 2006a]. Hazard surveillance should include the identification of work tasks and processes that involve the production and use of engineered nanoparticles, and should be viewed as one of the most critical components of any risk management program.

Discussion and Conclusions

Assessing the potential toxicity of engineered nanoparticles is at an early stage. A body of scientific evidence has accrued from toxicology studies on selected engineered nanoparticles and from epidemiology studies of individuals exposed to ultrafine nanoparticles suggests that some nanoscale particles may pose a health concern [Kuempel et al. 2007; Gwinn and Vallyathan 2006; Donaldson et al. 2006]. This evidence suggests that safety and health professionals should consider precautionary management approaches in workplaces where there is exposure to engineered nanoparticles [Schulte and Salamanca-Buentello 2007; NIOSH 2006a; Royal Society and Royal Academy of Engineering 2004; Borm et al. 2006; Holman et al. 2006; IRSST 2006] such as the implementation of occupational risk management programs. Such approaches are described in the NIOSH document *Approaches to Safe Nanotechnology: An Information Exchange with NIOSH* [NIOSH 2006a].

The current body of evidence about the possible health risks of occupational exposures to engineered nanoparticles is not sufficient to support the determination of specific medical screening to identify preclinical changes associated with exposure to engineered nanoparticles. No substantial link has been established between occupational exposure to engineered nanoparticles and adverse health effects. In addition, the toxicological research to date is insufficient to recommend such monitoring, the appropriate triggers for it, or components of it. As the volume of research on the potential health effects from exposure to engineered nanoparticles increases, continual reassessment will be needed to determine whether medical screening is warranted for workers who are producing or using engineered nanoparticles. NIOSH will continue to examine new research findings and update its recommendations on medical screening programs for workers exposed to nanoparticles. Appendix D provides a brief discussion concerning occupational health programs that include medical screening and might serve as a model for future reference for one or more engineered nanoparticles. Appendix E provides discussion highlighting details of instances where sufficient evidence to support recommendations for specific medical screening for workers exposed to engineered nanoparticles is lacking.

At this time, only a few types of engineered nanoparticles have been studied, and a clear and consistent picture of the relevant endpoints for workers has not yet emerged. Various physicochemical parameters of nanoparticles (e.g., composition, size, shape, surface characteristics, charge, functional groups, crystal structure, and solubility) appear to affect toxicity [Oberdörster et al. 2005a; Borm et al. 2006; Warheit et al. 2007b; IRSST 2006]. It is not known whether size is the overriding parameter, though most studies show that size appears to be the major factor in enhancing the toxicity of engineered nanoparticles compared with the toxicity of larger particles of the same composition. Results from a limited number of experimental animal studies with engineered nanoparticles indicate the potential for respiratory and circulatory effects [Aitken et al. 2004; Borm et al. 2006; ASCC 2006; IRRST 2006]; however, it is not clear which effects are most critical, whether they are dose-dependent, and whether these effects are relevant to human exposure. Additional studies are needed to determine the biological significance of different physicochemical parameters and whether these parameters can be used to predict the potential toxicity of other untested engineered nanoparticles.

When occupational health surveillance is being established, it is necessary to understand the relative, absolute, and population-attributable risks to workers who are handling engineered nanomaterials. This includes understanding the hazard as well as the extent of exposure and ultimately the risk. Limited information is available on these topics, but exposures may be generally low relative to the airborne exposures of the same material in larger but respirable particle sizes. The level of risk resulting from lower exposures to nanomaterials is unknown. Ultimately, epidemiological studies of exposed workers will be needed to help assess exposure-response relationships. Although such studies are difficult to conduct, they are more likely than medical screening to clarify the relationship between exposure and adverse effects at this time.

Finally, there is not yet enough research to make categorical determinations of the hazards based on combinations of physicochemical factors [ASCC 2006; Aitken et al. 2004]. Although preliminary studies indicate that while specific medical screening may be warranted in the fu-

ture, insufficient information is now available to make any recommendations beyond hazard surveillance. NIOSH will continue to assess the scientific evidence and periodically update the guidance on medical screening.

Recommendations

Continued *in vivo* and *in vitro* toxicological research is needed to identify potential health endpoints related to occupational exposure to engineered nanoparticles. Epidemiological studies of exposed workers will be needed to establish associations between exposures to engineered nanoparticles and adverse health effects and to assess other potential exposure-response relationships. Research is needed to assess various candidate biological markers that may ultimately be used in medical screening, including molecular markers [Schulte 2005]. This research is needed to assess sensitivity, specificity, and predictive value of biomarkers and clinical tests that might be developed and used to screen workers' health. Determining sufficient positive predictive value of a screening modality to detect adverse health effects early enough in the course of the disease to enable secondary prevention, is an important factor when considering medical screening efforts.

The following recommendations are provided for workplaces where workers may be exposed to engineered nanoparticles during the course of their work.

Take prudent measures to control exposures to engineered nanoparticles.

A prudent approach to controlling exposures to engineered nanoparticles has been described in the NIOSH draft document *Approaches to*

Safe Nanotechnology: An Information Exchange with NIOSH [NIOSH 2006a].

Conduct hazard surveillance as the basis for implementing controls.

To establish prudent measures for controlling exposure to engineered nanoparticles, it is important to identify which jobs or processes involve the production or use of engineered nanoparticles. Employers should identify and document the presence of engineered nanoparticles in their workplaces and the work tasks associated with them. This information will serve as the basis for applying various control measures [NIOSH 2006a]. Hazard surveillance programs should be designed to address some or all of the following questions:

— What exposure agents are found in the workplace?

— Are standardized, reliable, and practical methods available for measuring workers' exposures to the agents?

— What exposure metrics (e.g., mass, particle count, particle surface area) are most relevant to the most important health concerns?

— To what extent can specific exposures (e.g., nanoparticles) be linked to people?

— What actions have been taken to control potentially hazardous exposures?

— How effective are the controls (e.g., engineering)?

— Which agents affect the most workers?

— What jobs or industries are most likely to cause exposures to workers?

— What health effects are most likely related to these exposures?

— How are specific occupational exposures changing over time?

Continue use of established medical surveillance approaches.

Currently, there are many established uses for medical surveillance by employers and occupational health practitioners (see Section 3.3). These may pertain to workers exposed to engineered nanoparticles, but they are not specifically focused on them. Employers should continue using these established approaches to collect data that may be informative in the future about whether there is an increase in the frequency of adverse health effects related to exposure to engineered nanoparticles. NIOSH continues to recommend occupational health surveillance as an important part of an effective risk management program. Lack of evidence for recommending medical screening for workers potentially exposed to engineered nanoparticles should not stop employers who want to take additional precautions, including medical screening, beyond those already established. However, it is important to note that nonspecific medical testing can have negative consequences such as adverse effects resulting from tests (e.g., radiation from chest radiographs), creating unnecessary anxiety in workers and employers from false-positive screening tests, and the economic ramifications of additional diagnostic evaluations [Nasterlack et al. 2007; Schulte 2005; Marcus et al. 2006].

NIOSH will continue to evaluate the usefulness of establishing exposure registries in workplaces were there is potential exposure to engineered nanoparticles. As the understanding of occupational exposure to engineered nanoparticles increases, the development of exposure registries may be needed to form the basis for future epidemiologic research (Appendix F). Such registries probably need to cover workers from numerous companies to reflect the diversity of exposures, to account for the small number of workers exposed at a given site, and to assess chronic health effects.

References

Aitken RJ, Creely KS, Tran CL [2004]. Nanoparticles: an occupational hygiene review. United Kingdom: Health and Safety Executive. Research Report 274. HSE Books, Norwich UK.

Americans with Disabilities Act (ADA) [1990]. 42 USC 12101–12213.

Antonini [2003]. Health effects of welding. Crit Rev Toxicol 33(10):51–103.

Ashford NA, Spadafor CJ, Hattis DB, Caldart [1990]. Monitoring the worker for exposure and disease. Baltimore, MD: The Johns Hopkins University Press.

ASCC (Australian Safety and Compensation Council) [2006]. A review of the potential occupational safety and health implications of nanotechnology [http://www.ascc.gov.au/ascc/HealthSafety/EmergingIssues/Nanotechnology].

Baker EL, Matte TP [2005]. Occupational health surveillance. In: Rosenstock L, Cullen E, Brodkin R (eds). Textbook of clinical occupational and environmental medicine [http://www.osha.gov/SLTC/medicalsurveillance/surveillance.html]. Philadelphia, PA: Elsevier Saunders Company.

Borak J, Woolf SH, Fields CA [2006]. Use of beryllium lymphocyte proliferation testing for screening of asymptomatic individuals in evidence-based assessment. J Occup Environ Med 48:937–947.

Borm PJA, Robbins D, Haubald S, Kuhlbusch T, Fissan H, Donaldson K, Sching R, Stone V, Kreyling W, Lademann J, Krutmann J, Warheit D, Oberdörster E [2006]. The potential risks of nanomaterials: review carried out for ECETOC. Particle Fibre Toxicol 3(11). doi:10.1186/1743–8977–3–11.

Brown, DM, Wilson MR, MacNee W, Stone V, Donaldson K [2001]. Size-dependent proinflammatory effects of ultrafine polystyrene particles: a role for surface area and oxidative stress in the enhanced activity of ultrafines. Toxicol Appl Pharmacol 175(3):191–199.

Castranova V [2000]. From coal mine dust to quartz: mechanisms of pulmonary pathogenicity. Inhal Toxicol 12(Suppl.3):7–14.

CFR. Code of Federal Regulations. Washington, DC: U.S. Government Printing Office, Office of the Federal Register.

Dick CAJ, Brown DM, Donaldson K, Stone V [2003]. The role of free radicals in the toxic and inflammatory effects of four different ultrafine particle types. Inhal Toxicol 15(1):39–52.

Dockery DW, Pope CA, Xu X, Spengler JD, Ware JH, Fay ME, Ferris BG, Speizer BE [1993]. An association between air pollution and mortality in six U.S. cities. N Eng J Med 329(24):1753–1759.

Donaldson K, Tran L, Jimenez LA, Duffin R, Newby DE, Mills N, MacNee W, Stone V [2005]. Combustion-derived nanoparticles: a review of their toxicology following inhalation exposure. Part Fibre Toxicol 2:10–14.

Donaldson K, Aitken R, Tran L, Stone V, Duffin R, Forrest G, Alexander A [2006]. Carbon nanotubes; a review of their properties in relation to pulmonary toxicology and workplace safety. Toxicol Sci 92(1):5–22.

Duffin R, Tran L, Brown D, et al. [2007]. Proinflammogenic effects of low-toxicity and metal nanoparticles in vivo and in vitro: highlighting the role of particle surface area and surface reactivity. Inhal Toxicol 19:849–56.

Elder A, Gelein R, Silva V, Feikert T, Opanashuk L, Carter J, Potter R, Maynard A, Ito Y, Finkelstein J, Oberdörster G [2006]. Translocation of inhaled ultrafine manganese oxide particles to the central nervous system. Environ. Health Perspect. *114*(8):1172–1178.

Froines J, Wegman D, Eisen E [1989]. Hazard surveillance in occupational disease. Am J Public Health *79* (Suppl):26–31.

Garshick E, Laden F, Hart JE, Rosner B, Smith TJ, Dockery DW, Speizer FE [2004]. Lung cancer in railroad workers exposed to diesel exhaust. Environ Health Perspect *112*(15):1539–1543.

Gwinn MR, Vallyathan V [2006]. Nanoparticles: health effects pros and cons. Environ Health Perspect *114*(12):1818–25.

Halperin WE, Ratcliffe JM, Frazier JM, Wilson L, Becker SP, Schulte P [1986]. Medical screening in the workplace: proposed principles. J Occup Med *28* (8):522–547.

Halperin WE [1996]. The role of surveillance in the hierarchy of prevention. Am J Ind Med *29*(8):321–323.

Harber P, Conlon C, McCunney RJ [2003]. Occupational medical surveillance. In: McCunney RJ, (ed). A practical approach to occupational and environmental medicine. Philadelphia, PA: Lippincott, Williams, and Wilkins.

Helland A, Wick P, Koehler A [2007]. Reviewing the environmental and human health knowledge base of carbon nanotubes. Environ Health Perspect. *115*(8):1125–31.

Höhr D, Steinfartz Y, Schins RPF, Knaapen AM, Martra G, Fubini B, Borm P [2002]. The surface area rather than the surface coating determines the acute inflammatory response after instillation of fine and ultrafine TiO2 in the rat. Int J Environ Health *205*:239–244.

Holman MW [2006]. Nanotech environmental, health, and safety risks: action needed. In: Health and Nanotechnology: economics, societal, and institutional impact perspectives on the future of science and technology. Report from a conference convened with cooperation of the United States Department of State and the European Commission. Varenna, Italy, May 21–23.

IARC [2006]. Titanium dioxide. Lyon, France: International Agency for Research on Cancer [monographs.iarc.fr/ENG/Meetings/93-titaniumdioxide.pdf].

Ibald-Mulli A, Wichmann HE, Kreyling W, Peters A [2002]. Epidemiological evidence on health effects of ultrafine particles. J Aerosol Med Depos *15*(2):189–201.

ILO [1998]. Technical and ethical guidelines for workers' health surveillance (OSH No. 72). Geneva: International Labour Office.

IRSST [2006]. Nanoparticles: actual knowledge about occupational health and safety risks and prevention measures. Montreal, Canada: Institut de Recherche Robert-Sauvé en Santé et en Sécurité du Travail, R–470

Kreyling WG, Semmler-Behnke M, Moller W [2006]. Ultrafine particle-lung interactions: does size matter? J Aerosol Med *19*(1):74–83.

Kuempel ED, Geraci CL, Schulte PA [2007]. Risk assessment and research needs for nanomaterials: an examination of data and information from current studies. In: Simeonova PP et al. (eds). Nanotechnology-toxicological issues and environmental safety, Springer Publishing Company, pp. 119–145.

Lam CW, James JT, McCluskey R, Hunter RL [2004]. Pulmonary toxicity of single-wall carbon nanotubes in mice 7 and 90 days after intratracheal instillation. Toxicol Sci 77(1):126–134.

Lam CW, Jame JT, McCluskey R et.al. [2006]. A review of carbon nanotube toxicity and assessment of potential occupational and environmental risks. Crit Rev Toxicol 36:189–217.

Li Z, Hulderman T, Salmen R, Leonard SS, Young S-H, Shvedova A, Luster MI, Simeonova P [2007]. Cardiovascular effects of pulmonary exposure to single-wall carbon nanotubes. Environ Health Persp 115:377-382.

Lison, D, Lardot C, Huaux F, Zanetti G, Fubini B [1997]. Influence of particle surface area on the toxicity of insoluble manganese dioxide dusts. Arch Toxicol 71(12):725–729.

Marcus PM, Bergstalh EJ, Zweig MH, Harris A, Offord KP, Fontana RS [2006]. Extended lung cancer incidence follow-up in the Mayo lung project. J Natl Cancer Inst 7(98):748–56.

Maynard AD, Kuempel ED [2005]. Airborne nanostructural particles and occupational health. J Nanoparticles Res 7:587–614.

Maynard AD [2007]. Nanotechnology: the next big thing or much ado about nothing. Ann Occ Hyg 51(1):1–12.

Nasterlack M, Zober A, Oberlinner C [2007]. Considerations on occupational medical surveillance in employees handling nano-particles. Int Occ Environ Health. DOI 10.1007//S00420–0245–5.

NCI [2007]. Lung cancer screening (PDQ) [http://www.cancer.gov/cancertopics/pdq/screening/lung/patient/], p. 3.

Nel A, Xia T, Madlen L, Li W [2006]. Toxic potential of materials at the nanolevel. Science 311:622–627.

NIOSH [1998]. Criteria for a recommended standard: Occupational exposure to metal-working fluids. Cincinnati, OH: Department of Health and Human Services, Centers for Disease Control and Prevention, National Institute for Occupational Safety and Health, DHHS (NIOSH) Publication No. 1998–102.

NIOSH [2006a]. Approaches to safe nano-technology: an information exchange with NIOSH. Cincinnati, OH: Department of Health and Human Services, Centers for Disease Control and Prevention, National Institute for Occupational Safety and Health. DHHS (NIOSH) [www.cdc.gov/niosh/topics/nanotech/nano_exchange.html].

NIOSH [2006b]. Criteria for a recommended standard: occupational exposure to refractory ceramic fibers. Cincinnati, OH: Department of Health and Human Services, Centers for Disease Control and Prevention, National Institute for Occupational Safety and Health, DHHS (NIOSH) Publication No. 2006–125.

Nurkiewicz TR, Porter DW, Barger M, Rao KM, Marvar PJ, Hubbs AF, Castranova V, Boegehold MA [2006]. Systemic microvascular dysfunction and inflammation after pulmonary particulate matter exposure. Environ Health Perspect 114(3):412–9.

Nurkiewicz TR, Porter DW, Hubbs AF, Cumpston JL, Chen BT, Frazier DG, Castranova V. Nanoparticles inhalation augments particle-dependent systemic microvascular dysfunction. Particle and Fibre Toxicology [In Press].

Oberdörster G, Ferin J, Lehnert BE [1994]. Correlation between particle-size, in-vivo particle persistence, and lung injury. Environ Health Perspect *102*(S5):173–179.

Oberdörster G, Sharp Z, Atudorei V, Elder A, Gelein R, Lunts A, Kreyling W, Cox C [2002]. Extrapulmonary translocation of ultrafine carbon particles following whole-body inhalation exposure of rats. J Toxicol Environ Health *65* Part A (20):1531–1543.

Oberdörster G, Maynard A, Donaldson et al. [2005a]. Principles for characterizing the potential human health effects from exposure to nanomaterials: elements of a screening strategy. Particle and Fiber Toxicology *2*(8):1–113.

Oberdörster G, Oberdörster E, Oberdörster J [2005b]. Nanotoxicology—an emerging discipline involving studies of ultrafine particles. Environ Health Perspect *113*(7):823–839.

Occupational Safety and Health Administration (OSHA). Metalworking Fluids: Safety and Health Best Practices Manual. Available at: www.osha.gov/SLTC/metal workingfluids/metalworkingfluids_manual. html, last updated 3/2008.

Poland CA, Duffin R, Kinloch I, Maynard A, Wallace WAH, Seaton A, Stone V, Brown S, MacNee W, Donaldson K [2008]. Carbon nanotubes introduced into the abdominal cavity of mice show asbestos-like pathogenicity in a pilot study. Nature Nanotechnology. Published online: 20 May 2008; doi:10.1038/nnano.2008.111.

Pope CA, Burnett RT, Thurston GD, Thun MJ, Calle, EE, Krewski D, Godleski JJ [2004]. Cardiovascular mortality and long-term exposure to particulate air pollution: epidemiological evidence of general pathophysiological pathways of disease. Circulation *109*(1):71–74.

Radomski A, Juraz P, Alonso-Escolano D, Drews M, Morandi M, Malinsk T, et al. [2005]. Nanoparticle-induced platelet aggregation and vascular thrombosis. Br J Pharmacol *146*(6):882–893.

Royal Society and Royal Academy of Engineering [2004]. Nanoscience and nanotechnologies: opportunities and uncertainties. London: The Academy.

Ruckerl R, Ibald-Mulli A, Koenig W, Schneider A, Woelke G, Cyrys J, Heinrich J, Marder V, Frampton M, Wichmann HE, Peters A [2006]. Air pollution and markers of inflammation and coagulation in patients with coronary heart disease. Am J Respir Crit Care Med *173*(4):432–441.

Schulte PA, Kaye WE [1988]. Exposure registries. Arch Environ Health *43*:155–161.

Schulte PA [2005]. The use of biomarkers in surveillance, medical screening, and intervention. Mutation Res *592*(1–2):155–163.

Schulte PA, Salamanca-Buentello F [2007]. Ethical and scientific issues of nanotechnology in the workplace. Env Health Perspectives *115*:5–12.

Shvedova AA, Kisin ER, Mercer R, Murray AR, Johnson VJ, Potapovich A, et al. [2005]. Unusual inflammatory and fibrogenic pulmonary responses to single-walled carbon nanotubes in mice. Am J Physiol Lung Cell Mol Physiol *289*:L698–L708.

Shevedova AA, Kisin E, Murray AR, Johnson V, Gorelik O, Arepalli S, Hubbs AF, Mercer RR, Stone S, Frazer D, Chen T, Deye G, Maynard A, Baron P, Mason R, Kadiiska M, Stadler K, Mouithys-Mickalad A, Castranova V, Kaagan VE [2008]. Inhalation of carbon nanotubes induces oxidative stress and cytokine response causing respiratory impairment and pulmonary fibrosis in mice. The Toxicologist *102*:A1497.

Steenland K, Deddens J, Stayner L [1998]. Diesel exhaust and lung cancer in the trucking industry: exposure-response analyses and risk assessment. Am J Med *34*(3):220–228.

Sundin DS, Frazier TM [1989]. Hazard surveillance at NIOSH. Am J Pub Health *79*(Suppl):32–37.

Takagi A, Hirose A, Nishimura T, Fukumori N, Ogata A, Ohashi N, Kitajima S, Kanno J [2008]. Induction of mesothelioma in p53+/- mouse by intraperitoneal application of multi-wall carbon nanotube. J Toxicol Sci. *33*(1):105–16.

Wagner GR, Fine LJ [2008]. Surveillance and health screening in occupational health. In: Maxcy-Rosenau-Last Public Health and Preventive Medicine. RB Wallace (ed.) 15th ed. New York: McGraw-Hill Medical Publishing, pp 759–793.

Warheit DB, Webb TR, Colvin VL, Reed KL, Sayes CM [2007a]. Pulmonary bioassay studies with nanoscale and fine-quartz particles in rats: toxicity is not dependent upon particle size but on surface characteristics. Toxicol Sci *95*(1):270–280.

Warheit DB, Webb TR, Reed KL, Frerichs S, Sayes CM [2007b]. Pulmonary toxicity study in rats with three forms of ultrafine-TiO_2 particles: differential response related to surface properties. Toxicology *230*(1):90–104.

Zhang QW, Kusaka Y, Sato K, Nakakuki K, Kohyama N, Donaldson K [1998]. Differences in the extent of inflammation caused by intratracheal exposure to three ultrafine metals: role of free radicals. J Toxicol Environ Health-Part A *53*(6):423–438.

Zhang Q, Kusaka Y, Donaldson K [2000]. Comparative pulmonary responses caused by exposure to standard cobalt and ultrafine cobalt. J Occup Health *42*:179–184.

Zhang Q, Kusaka Y, Zhu X, Sato K, Mo Y, Klutz T, Donaldson K [2003]. Comparative toxicity of standard nickel and ultrafine nickel in lung after intratracheal instillation. J Occup Health *45*:23–30.

APPENDIX A · Critical Aspects of an Occupational Medical Screening Program

Assessment of workplace hazards

Identification of target organ toxicities for each hazard

Selection of test for each "screenable health effect"

Development of action criteria

Standardization of data collection process

Performance of testing

Interpretation of test results

Test confirmation

Determination of work status

Notification

Diagnostic evaluation

Evaluation and control of exposure

Recordkeeping

[Baker and Matte 2005].

APPENDIX B · OSHA Standards That Include Requirements for Medical Surveillance

2-acetylaminofluorene

acrylonitrile

4-aminodiphenyl

inorganic arsenic

asbestos

benzene

benzidine

bis-chloromethyl ether

1,3–butadiene

coke oven emissions

cotton dust

dibromochloropropane

3.3'-dichlorobenzidine

4-dimethylaminoazobenzene

cadmium

occupational exposure to hazardous chemicals in the laboratories

ethylene oxide

ethyleneimine

formaldehyde

hazardous waste

lead

methyl chloromethyl ether

alpha-naphthylamine

beta-naphthylamine

methylene chloride

4-nitrobiphenyl

n-nitrosodimethylamine

beta-propriolactone

vinyl chloride

methylenedianiline

bloodborne pathogens

chromium (VI)

APPENDIX C · Hazards for Which NIOSH Has Recommended the Use of Medical Surveillance

NIOSH publication number	Title and date	NTIS stock number
76–195	Acetylene (1976)	PB 267068
77–112	Acrylamide (1976)	PB 273871
78–116	Acrylonitrile (1978)	PB 81–225617
77–151	Alkanes (C5-C8) (1977)	PB 273817
76–204	Allyl Chloride (1976)	PB 267071
74–136	Ammonia (1974)	PB 246699
78–216	Antimony (1978)	PB 81–226060
74–110	Arsenic, Inorganic (1974), (Revised 1975)	PB 228151
75–149	Arsenic, Inorganic (1975)	PB 246701
72–10267	Asbestos (1972)	PB 209510
77–169	Asbestos (Revised) (1976)	PB 273965
78–106	Asphalt Fumes (1977)	PB 277333
74–137	Benzene (1974)	PB 246700
*	Benzene (Revised) (1976)	PB 83–196196
77–166	Benzoyl Peroxide (1977)	PB 273819
78–182	Benzyl Chloride (1978)	PB 81–226698
72–10268	Beryllium (1972)	PB 210806
*	Beryllium (Revised) (1977)	PB 83–182378
	2-Butoxyethanol [See: Ethylene Glycol Monobutyl Ether]	
77–122	Boron Trifluoride (1976)	PB 274747
76–192	Cadmium (1976)	PB 274237
77–107	Carbaryl (1976)	PB 273801
78–204	Carbon Black (1978)	PB 81–225625
76–194	Carbon Dioxide (1976)	PB 266597

(continued)

NIOSH publication number	Title and date	NTIS stock number
77–156	Carbon Disulfide (1977)	PB 274199
73–11000	Carbon Monoxide (1972)	PB 212629
76–133	Carbon Tetrachloride (1975)	PB 250424
*	Carbon Tetrachloride (Revised) (1979)	PB 83–196436
76–170	Chlorine (1976)	PB 266367
75–114	Chloroform (1974)	PB 246695
*	Chloroform (Revised 1979)	PB 83–195856
77–210	Chloroprene (1977)	PB 274777
73–11021	Chromic Acid (1973) [Revised; see Chromium VI]	PB 222221
760–129	Chromium VI (1975)	PB 248595
78–191	Coal Gasification Plants (1978)	PB 80–164874
95–106	Coal Mine Dust	PB 96–191713
78–107	Coal Tar Products (1977)	PB 276917
82–107	Cobalt (1981)	PB 82–182031
73–11016	Coke Oven Emissions (1973)	PB 216167
80–106	Confined Spaces, Working in Construction [See: Excavations] (1979)	PB 80–183015
75–118	Cotton Dust (1974)	PB 246696
78–133	Cresol (1978)	PB 86–121092
77–108	Cyanide, Hydrogen and Cyanide Salts (1976)	PB 266230
78–115	Dibromochloropropane (1978) 1,2-Dichloroethane [See: Ethylene Dichloride]	PB 81–228728
96–104	2-Diethylaminoethanol (1996)	PB 96–197371
78–215	Diisocyanates (1978)	PB 81–226615
78–131	Dinitro-ortho-Cresol (1978)	PB 80–175870
77–226	Dioxane (1977)	PB 274810
76–128	Elevated Work Stations, Emergency Egress from (1975)	PB 248594
76–206	Epichlorohydrin (1976)	PB 81–227019
77–221	Ethylene Dibromide (1977)	PB 276621
76–139	Ethylene Dichloride (1976)	PB 85–178275
78–211	Ethylene Dichloride (1,2- Dichloroethane) (Revised) (1978)	PB 80–176092

(continued)

NIOSH publication number	Title and date	NTIS stock number
90–118	Ethylene Glycol Monobutyl Ether and Ethylene Glycol Monobutyl Ether Acetate (1991)	PB 91–173369
91–119	Ethylene Glycol Monomethyl Ether, Ethylene Glycol Monoethyl Ether, and Their Acetates	PB 92–167147
83–103	Excavations, Development of Draft Construction Safety Standards for, Volume 1 (1983)	PB 84–100569
*	Excavations, Development of Draft Construction Safety Standards for, Volume 2 (1983)	PB 83–233353
77–152	Fibrous Glass (1977)	PB 274195
76–103	Fluorides, Inorganic (1975)	PB 246692
77–193	Fluorocarbon Polymers, Decomposition Products of (1977)	PB 274727
77–126	Formaldehyde (1976)	PB 273805
85–116	Foundries (1985)	PB 86–213477
79–133	Furfuryl Alcohol (1979)	PB 80–176050
78–166	Glycidyl Ethers (1978)	PB 81–229700
83–126	Grain Elevators and Feed Mills (1983)	PB 83–138537
89–106	Hand-Arm Vibration (1989)	PB 90–168048
83–125	Guidelines for Controlling Hazardous Energy During Maintenance and Servicing (1983)	PB 84–199934
72–10269	Hot Environments (1972)	PB 210794
86–113	Hot Environments (Revised 1986)	PB 86–219508
78–172	Hydrazines (1978)	PB 81–225690
	Hydrogen Cyanide [See: Cyanide, Hydrogen and Cyanide Salts]	
76–143	Hydrogen Fluoride (1976)	PB 81–226516
77–158	Hydrogen Sulfide (1977)	PB 274196
78–155	Hydroquinone (1978)	PB 81–226508
75–126	Identification System for Occupationally Hazardous Materials (1974)	PB 246698
76–142	Isopropyl Alcohol (1976)	PB 273873
*	Kepone (1976)	PB 83–196170
78–173	Ketones (1978) Labeling [See: Identification System for Occupationally Hazardous Materials]	PB 80–176076
73–11010	Lead, Inorganic (1972)	PB 214265
78–158	Lead, Inorganic (Revised) (1978)	PB 81–225278
	Lockout/Tagout [See: Hazardous Energy]	

(continued)

NIOSH publication number	Title and date	NTIS stock number
76–188	Logging from Felling to First Haul (1976)	PB 266411
76–205	Malathion (1976)	PB 267070
73–11024	Mercury, Inorganic (1973)	PB 222223
76–148	Methyl Alcohol (1976)	PB 273806
	Methyl Chloroform [See: 1,1,1-Trichloroethane]	
77–106	Methyl Parathion (1976)	PB 274191
76–138	Methylene Chloride (1976)	PB 81–227027
98–102	Metalworking Fluids (1998)	PB 99–133910
77–164	Nickel, Inorganic (1977)	PB 274201
76–141	Nitric Acid (1976)	PB 81–227217
78–212	Nitriles (1978)	PB 81–225534
76–149	Nitrogen, Oxides of (1976)	PB 81–226995
78–167	Nitroglycerin and Ethylene Glycol Dinitrate (1978)	PB 81–225526
73–11001	Noise (1972)	PB 213463
2006–123	Occupational Exposure to Refractory Ceramic Fibers	
98–126	Occupational Noise Exposure	PB 98–173–735
83–127	Oil and Gas Well Drilling (1983)	PB 84–242528
77–115	Organotin Compounds (1976)	PB 274766
84–115	Paint and Allied Coating Products (1984)	PB 85–178978
76–190	Parathion (1976)	PB 274192
	Perchloroethylene [See: Tetrachloroethylene]	
78–174	Pesticides, Manufacture and Formulation	PB 81–227001
76–196	Phenol (1976)	PB 266495
76–137	Phosgene (1976)	PB 267514
77–225	Polychlorinated Biphenyls (1977)	PB 276849
84–103	Precast Concrete Products Industry (1984)	PB 85–220051
88–101	Radon Progeny in Underground Mines (1988)	PB 88–173455
77–192	Refined Petroleum Solvents (1977)	PB 85–178267
2006–123	Refractory Ceramic Fibers (2006)	PB 2006–1123003
75–120	Silica, Crystalline (1974)	PB 246697
76–105	Sodium Hydroxide (1975)	PB 246694
83–119	Styrene (1983)	PB 84–148295
74–111	Sulfur Dioxide (1974)	PB 228152
*	Sulfur Dioxide (Revised) (1977)	PB 83–182485
74–128	Sulfuric Acid (1974)	PB 233098

(continued)

NIOSH publication number	Title and date	NTIS stock number
77–121	1,1,2,2-Tetrachloroethane (1976)	PB 273802
76–185	Tetrachloroethylene (Perchloroethylene) (1976)	PB 266583
78–213	Thiols: N-Alkane Mono, Cyclohexane, and Benzene (1978)	PB 81–225609
78–179	o-Tolidine (1978)	PB 81–227084
73–11023	Toluene (1973)	PB 222219
73–11022	Toluene Diisocyanate (1973) [Revised; See: Diisocyanates]	PB 222220
76–184	1,1,1-Trichloroethane (Methyl Chloroform) (1976)	PB 267069
73–11025	Trichloroethylene (1973)	PB 222222
77–127	Tungsten and Cemented Tungsten Carbide (1977)	PB 275594
73–11009	Ultraviolet Radiation (1972)	PB 214268
77–222	Vanadium (1977)	PB 81–225658
78–205	Vinyl Acetate (1978)	PB 80–176993
*	Vinyl Chloride (1974)	PB 246691
*	Vinyl Halides (1979)	PB 84–125699
77–140	Waste Anesthetic Gases and Vapors (1977)	PB 274238
88–110	Welding, Brazing, and Thermal Cutting (1988)	PB 88–231774
75–168	Xylene (1975)	PB 246702
76–104	Zinc Oxide (1975)	PB 246693

*Denotes the absence of a publication number or that recommendations were provided in testimony by NIOSH to the U.S. Department of Labor.

NTIS [National Technical Information Service] Web site: http://www.ntis.gov

APPENDIX D · Discussion of Occupational Health Surveillance Programs with Medical Screening

Occupational health surveillance programs exist that may be useful as models on which to base future efforts in the management of occupational exposures to one or more engineered nanoparticles together with any potential health risk(s) related to those exposures.

Occupational exposures to metalworking fluids (MWF) have been implicated in health problems including a variety of dermatologic and respiratory health conditions. In the *NIOSH Criteria for a Recommended Standard: Occupational Exposure to Metal Working Fluids* [NIOSH 1998], medical monitoring (screening) is recommended by NIOSH as part of a complete MWF safety and health program. Similarly, in the Safety and Health Best Practices Manual for Metalworking Fluids [OSHA, 2008], OSHA recommends a model for a medical monitoring (screening) program and provides information on implementation. These recommended programs provide examples of how appropriate occupational health surveillance principles may be applied toward prevention and control of occupational exposures and associated health risks. Although there are still scientific uncertainties related to occupational exposures to MWF that require further research, the recommendations for these medical monitoring programs are based on evaluation of extensive data concerning exposures, health effects, and exposure-health effect relationships. As noted in the *NIOSH Criteria Document for a Recommended Standard*, these recommendations concerning MWF are made with the expectation that they will prevent or greatly reduce the risk of adverse health effects in exposed workers. Gathering exposure and health effect data related to occupational exposure to engineered nanoparticles will be essential when formulating medical screening programs for workers exposed to engineered nanoparticles.

APPENDIX E · Examples of Limitations in the Evidence Base for Specific Medical Screening of Workers Exposed to Engineered Nanoparticles

Key among the criteria for recommending specific medical screening of workers exposed to engineered nanoparticles include determining whether the substance in question is a hazard and whether the disease to be averted is sufficiently common in the worker population to justify routine screening [Nasterlack et al. 2007; Borak et al. 2006; Halperin et al. 1986]. For engineered nanoparticles, there is insufficient evidence for a definitive hazard determination. Only a small number of the myriad types of engineered nanoparticles have undergone experimental animal inhalation testing, and no broad categories of physicochemical risk factors have been identified to allow for projecting hazards across particle types. No chronic inhalation studies of engineered nanoparticles have been conducted to date. The existence of a few short-term inhalation studies on carbon nanotubes and nanoscale metal oxides is not adequate to identify what disease endpoints to assess in medical screening. There is also insufficient information available regarding the absolute, relative or population-attributable risks associated with nanoparticle exposures [Nasterlack et al. 2007].

Examples of the issues in determining the rationale for recommending medical screening for workers potentially exposed to engineered nanoparticles are described as follows.

Single-Walled Carbon Nanotubes (SWCNTs)

Intratracheal (IT) exposure to SWCNTs has been associated with interstitial fibrosis in the rat (Lam et al. 2004]. Aspiration of purified SWCNTs caused rapid and progressive interstitial fibrosis in mice [Shvedova et al. 2005]. NIOSH has also shown that inhalation of SWCNTs cause interstitial fibrosis [Shvedova et al. 2008]. The problem is that purified SWCNTs are not redox reactive and the interstitial fibrosis is not driven by oxidant generation and inflammation. Therefore, measurement of markers of oxidant stress or inflammation in humans would not be predictive. If interstitial lung disease was considered the health endpoint of concern, monitoring of the carbon monoxide diffusion capacity of the lung could be performed noninvasively. A significant decline in diffusion would indicate a loss of alveolar-capillary gas exchange and suggest early signs of pre-clinical disease. Unfortunately, virtually no published data exist on occupational exposure concentrations for working in SWCNT operations. Consequently, there is too little information available at this time to verify disease endpoints. There is also too little information available on exposure in general and ultimately the risk to workers who handle these materials.

Nanoscale Metal Oxides

Pulmonary exposure to nanoscale metal oxides such as titanium dioxide (TiO_2) have been shown in rat models to cause pulmonary inflammation [Oberdörster et al. 2005] and to inhibit the ability of the systemic microvasculature to respond to dilators [Nurkiewicz et al.2006; Nurkiewicz et al. in press] after IT or inhalation exposures. Ultrafine (nanoscale) TiO_2 has been shown to be more potent in causing these effects than fine TiO_2 on an equivalent mass basis. These effects have been associated with oxidant stress and induction of inflammatory mediators. Therefore, markers of oxidant stress and inflammation could be considered as early indicators of human exposure or response. Oxidant stress markers have been suggested as markers of toxicity to metal oxide nanoparticles as a class [Nel et al. 2006]. Examples of such markers would be nitrous oxide or isoprostanes in exhaled breath or blood markers of oxidant stress. However, the utility of these markers for screening workers exposed to engineered nanoparticles has not been demonstrated. In addition, some research shows that nanoscale TiO_2 is linked to cancer of the lung and the International Agency for Research on Cancer (IARC) has categorized titanium dioxide as a possible carcinogen to humans [IARC 2006]. Nonetheless, no evidence clearly demonstrates that medical screening of asymptomatic workers exposed to lung carcinogens decreases the chance of dying from cancer (NCI 2007; Marcus et al. 2006).

Nanoscale Cadmium

Cadmium is a substance that has medical screening recommendations for workers exposed in order to prevent or assess lung and kidney toxicity (see Appendices B and C). At a minimum, these recommendations should pertain to nanoscale cadmium (e.g., such as that used in the production of quantum dots). Medical screening is typically triggered by the airborne concentration of the substance in the workplace (e.g., the "action level" concentration). An action level is some fraction, usually 50%, of an occupational exposure limit (OEL). Whether the action level concentration recommended for nonnanoscale cadmium particles is adequate for nanoscale cadmium is unknown. Workplaces with engineered nanoparticles of materials addressed by current OSHA standards are subject to the requirements of those standards, including the requirements for medical surveillance.

APPENDIX F · Exposure Registries

Exposure registries are useful tools for surveillance of new or perceived hazards. A registry provides a structured and orderly approach to handling the problem of identifying and maintaining communication with workers exposed to hazardous substances [Schulte and Kaye 1988]. An exposure registry is the enrollment of persons exposed or likely to have been exposed to occupational or environmental hazards; the registry may include how these groups are managed with regard to primary or secondary preventive efforts. In occupational situations, company employee rosters are de facto registries; however, they may not address employees who leave a company. Moreover, for a new cross-cutting technology such as nanotechnology, the registry could enroll persons from various companies. Generally, exposure registries are developed and maintained by government entities, but there are examples of private-sector registries related to exposure to commercial products.

The purposes and functions of exposure registries may be summarized as follows:

- Delineate a population at risk

- Follow cohort to ascertain exposure-disease associations

- Follow cohort to ensure the institution of appropriate primary and secondary prevention and medical surveillance

- Follow cohort to allow for appropriate social, legal, and economic support

- Demonstrate societal concern for the cohort and provide a base for political action relevant to the exposure

- Notify a cohort of an exposure, preventive measures, or therapeutic advances that were not understood or known at the time the registry was established

Various issues should be addressed when considering development of exposure registries. These include the term of the registry, needs of registrants, confidentially of information, cost of maintaining the registry, and the potential impact of the registry on workers and companies.

Registries are essentially a collection of individual worker information over time with at least a preliminary plan for analysis. Data collected in registries may be subject to limitations. Exposure registries are not always useful in etiologic research. For diseases with low prevalence following low-level exposures, exposure registries are not very effective tools because (1) exposure classification is often difficult, (2) the statistical power of prospective studies is low, and (3) the time period of the study may be impractically long. Moreover, changes in exposures experienced by registry participants over time may complicate the ability to establish clear exposure-disease relationships.

Exposure registries may provide opportunities to determine the exposure-disease association and risk. Also, when practical prospective studies can be designed, registries can be used to establish hypotheses. Many questions arise when considering an exposure registry for etiologic research, including:

- How can exposed persons be adequately differentiated from nonexposed persons?

- What group could serve as a comparison group so that the disease experience of the exposed group can be evaluated?
- How long should the group be followed?

Although exposure registries are useful tools to assist populations potentially at risk, their utility for workers exposed to engineered nanoparticles needs further evaluation.

Notes

Notes